Octopuses

by Grace Hansen

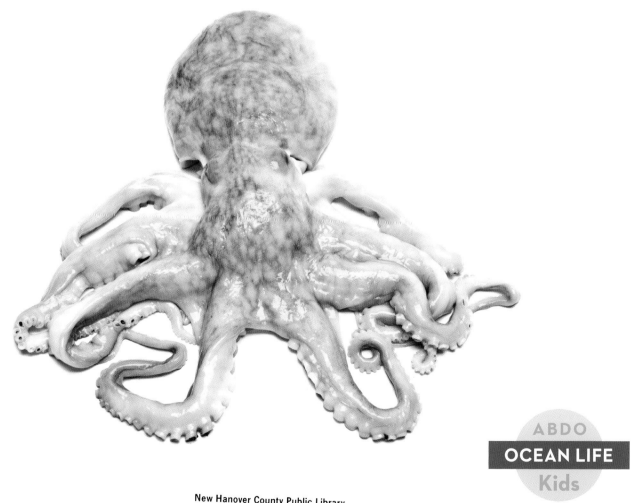

ABDO
OCEAN LIFE
Kids

abdopublishing.com

Published by Abdo Kids, a division of ABDO, PO Box 398166, Minneapolis, Minnesota 55439.

Copyright © 2015 by Abdo Consulting Group, Inc. International copyrights reserved in all countries. No part of this book may be reproduced in any form without written permission from the publisher.

Printed in the United States of America, North Mankato, Minnesota.

102014
012015

THIS BOOK CONTAINS RECYCLED MATERIALS

Photo Credits: iStock, Seapics.com, Shutterstock, Thinkstock
Production Contributors: Teddy Borth, Jennie Forsberg, Grace Hansen
Design Contributors: Laura Rask, Dorothy Toth

Library of Congress Control Number: 2014943717
Cataloging-in-Publication Data
Hansen, Grace.
  Octopuses / Grace Hansen.
   p. cm. -- (Ocean life)
ISBN 978-1-62970-710-5 (lib. bdg.)
Includes index.
1. Octopuses--Juvenile literature.   I. Title.
594/.56--dc23
                   2014943717

Table of Contents

Octopuses 4

Body Parts 8

Eating 14

Eggs 16

Baby Octopuses 18

More Facts 22

Glossary 23

Index 24

Abdo Kids Code 24

Octopuses

Octopuses live in the world's oceans. They live near rocky **coasts**. They also live in open waters.

Octopuses come in all sizes. They usually weigh between 6 and 10 pounds (2.7–4.5 kg). They are usually between 12 and 36 inches (30–91 cm).

Body Parts

All octopuses have a head.

They have eight arms.

Each arm has suckers.

9

Arms are used for moving and eating. Suckers hold onto **prey**.

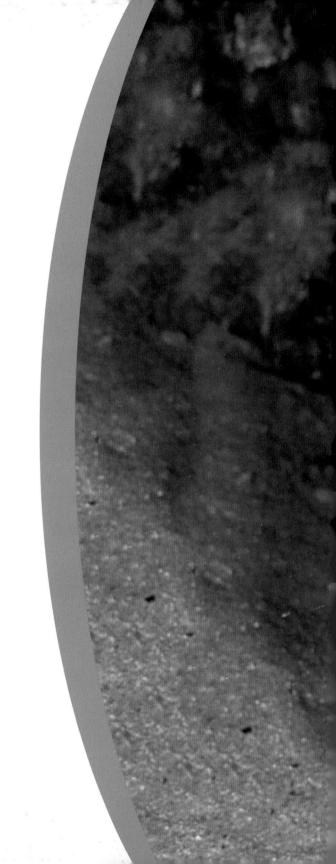

11

An octopus has a **beak**. It is the only hard part on its body. The beak tears and eats food.

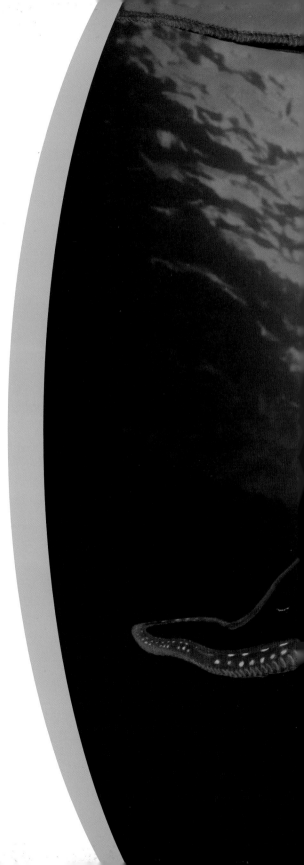

- - - - - - - - - - - beak

Eating

Octopuses kill their food with **venom**. This makes eating easier.

15

Eggs

Most females lay eggs once. They lay between 50 and 100,000 eggs. They lay them in rocks or holes.

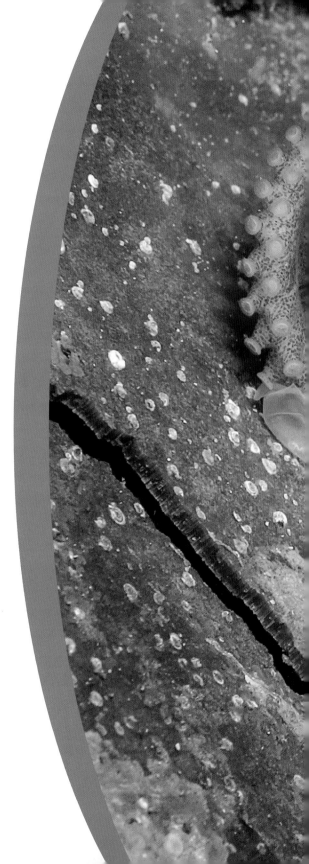

17

Baby Octopuses

The female stays with her eggs. It can take 2 to 14 months for them to hatch. Babies are on their own after they hatch.

19

Baby octopuses grow fast. Some will live on the ocean floor. Others will find small spaces to call home.

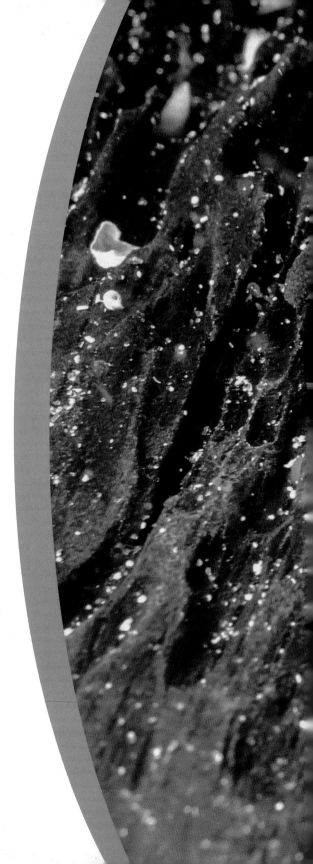

21

More Facts

- There are over 200 kinds of octopus. About half of them live in caves or cracks in rocks. The other half live on the ocean floor.

- Some octopuses can blend in with their surroundings. This is to keep them safe from **predators**. If they are spotted and feel scared, they let out a cloud of ink. Then they swim quickly away.

- Octopuses love to eat fish, lobsters, crabs, and clams.

Glossary

beak – a hard mouthpart that sticks out.

coast – land near an ocean.

predator – an animal that kills other animals for food.

prey – an animal hunted or killed for food.

venom – a poison made by some animals.

Index

babies 18, 20

body parts 8, 10, 12

eating 10, 12, 14

eggs 16, 18

food 10, 12, 14

habitat 4, 20

hunting 10, 14

ocean 4

size 6

abdokids.com

Use this code to log on to abdokids.com and access crafts, games, videos, and more!

Abdo Kids Code: OOK7105

ML 11-15